Sponges
Science Under The Sea

Lynn M. Stone

Rourke
Publishing LLC
Publishing LLC
Vero Beach, Florida 32964

PHOTO CREDITS: Cover, p. 10, 12, 20 © Marty Snyderman; p. 7, 8, 13, 15, 16,
19 © Brandon Cole; title page, p. 4 © James H. Carmichael.

EDITOR: Frank Sloan

COVER DESIGN: Nicola Stratford

Cover Photo: *A yellow tube sponge*

Library of Congress Cataloging-in-Publication Data

Stone, Lynn M.
  Sponges / Lynn M. Stone.
     p. cm. — (Science under the sea)
Summary: Describes the physical characteristics, behavior, and habitat
of these plantlike sea animals.
Includes bibliographical references (p.   ).
  ISBN 1-58952-322-9 (hardcover)
  1. Sponges—Juvenile literature. [1. Sponges.] I. Title.
  QL371.4 .S76 2002
  593.4—dc21
                                                    2002005133

Printed in the USA

CG/CG

Table of Contents

Animals That Look Like Plants

For many years, people didn't know that sea sponges were animals! Until the 1800s, many scientists thought sponges were plants.

Early scientists had been fooled because sponges look like plants. And most sponges stay in one place, as plants do. But plants make their own food. Sponges, like other animals, eat food that they bring into their bodies.

Scientists once believed that sponges were plants because they looked so much like plants.

Sea Animals

Most of the 5,000 **species** of sponges are sea animals. Only about 150 species live in fresh water.

Sponges live throughout the oceans. Sponges have been found in water that is more than 4 miles (6.4 kilometers) deep.

Like most sponges, this stove-pipe sponge in the Caribbean Sea lives in warm water.

Living in Warm Seas

Most sea, or **marine**, sponges live in warm seas. Divers find colorful sponges on **coral reefs**. People also find sponges that storms have washed onto beaches, especially in the southern United States.

A sponge generally attaches itself onto a rock, reef, or plant. Almost always, the undersea object becomes the sponge's lifelong home.

A diver swims to a barrel sponge on a Caribbean coral reef. This sponge is 5 feet (1.5 meters) across.

Slow Motion

At least one kind of sponge can release itself and move away. Movement for a sponge is not like "moving" for a person. A sponge might travel half an inch (12.7 millimeters) in one day. It would take a sponge about 20 years to go from one end of a football field to the other.

This orange puffball sponge has attached itself to a rock.

Sponges, like this red cup sponge, are often named for the objects they look like.

A crab has the perfect disguise: a covering of living sponge!

Sponge Bodies

Sponges are simple animals. They have no bones, arms, legs, brains, eyes, or ears.

A sponge body may be any of a huge variety of forms. It may look like a vase, candleholder, tube, or fingers. Or it may have no real shape at all. Some sponges, like moss, simply grow over an object.

The cobalt sponge simply spreads over the hard object beneath it.

Water Pumps

The smallest sponges are less than 1 inch (2.5 centimeters) across. Large sponges can be more than 6 feet (1.8 meters) wide.

All sponges are basically water-pumping systems. A sponge draws water into its body through tiny **pores**. It pumps water and food waste back out through a larger opening. Some sponge bodies have a system of canals.

A diver looks at an elephant ear sponge, a type that can be up to 6 feet (1.8 meters) across.

Feeding a Sponge

Water brings tiny animals and plants into the sponge. A sponge **filters** them from the water for its food.

A sponge's holes and waterways make great hangouts for other marine animals. A sponge taken from the Gulf of Mexico had more than 17,000 little animals in it!

Some animals eat sponges. These creatures include some sea turtles, snails, sea stars, and fish.

A rockfish hides in the opening of a giant cloud sponge in the cold North Pacific Ocean.

Sponges and People

Certain sea sponges have been used by people for hundreds of years. Sponges have been paint rollers, washcloths, table scrubbers, bandages, mops, drinking cups, and even body padding.

Sponges make some very strong **substances**. The fire sponges of the West Indies, for example, cause irritation of human skin.

Scientists are very interested in these substances. They may be used as ingredients of medicines for people.

The fire sponge can cause burn-like marks on human skin.

Many sponge bodies, however, are hard, crusty, or breakable. Only the types of sponges that make their skeletons of **spongin** are useful to people. In a few places, like Tarpon Springs, Florida, divers still harvest these bath sponges for home use. Most sponges used in homes today, however, were made in factories.

Glossary

coral reefs (KOR uhl REEFS) — huge, undersea limestone structures made from the skeletons of stony corals

filters (FILL turz) — to strain

marine (meh REEN) — from or having to do with the sea

pores (POURZ) — tiny openings through which air or water may pass

species (SPEE sheez) — within a group of closely related animals, one certain kind, such as a tube sponge

spongin (SPUN jin) — the substance that makes certain sponges soft and flexible

substances (SUB stan sez) — particular material or matter, such as blood or water

Index

Further Reading

Juster, Barbara. *Sponges Are Skeletons.* Harper Trophy, 1998
Llamas, Andreu. *Sponges: Filters of the Sea.* Gareth Stevens, 1997

Websites To Visit

Sponges: http://www.oceanoasis.org/fieldguide/sponges.html
Sponges: http://www.seasky.org/reeflife/sea2a.html

About The Author

Lynn Stone is the author of more than 400 children's nonfiction books. He is a talented natural history photographer as well. Lynn, a former teacher, travels worldwide to photograph wildlife in its natural habitat.